World of Bugs

FREAKY FLIES

By Greg Roza

Please visit our Web site, www.garethstevens.com. For a free color catalog of all our high-quality books, call toll free 1-800-542-2595 or fax 1-877-542-2596.

Library of Congress Cataloging-in-Publication Data

Roza, Greg.
Freaky flies / Greg Roza.
p. cm. — (World of bugs)
ISBN 978-1-4339-4596-0 (pbk.)
ISBN 978-1-4339-4597-7 (6-pack)
ISBN 978-1-4339-4595-3 (library binding)
1. Flies—Juvenile literature. I. Title.
QL533.2.R69 2011
595.77—dc22
2010022234

First Edition

Published in 2011 by
Gareth Stevens Publishing
111 East 14th Street, Suite 349
New York, NY 10003

Editor: Greg Roza
Designer: Christopher Logan

Photo credits: Cover, pp. 1, 3, 5, 7 (all), 9, 13, 15, 17, 23, 24 (housefly, legs, mosquito, wings) Shutterstock.com; p. 11 Eco Images/Universal Images Group/Getty Images; pp. 19, 24 (eggs) Stephen Dalton/Minden Pictures/Getty Images; p. 21 iStockphoto/Thinkstock.

Printed in the United States of America

CPSIA compliance information: Batch #CW11GS: For further information contact Gareth Stevens, New York, New York at 1-800-542-2595.

FREAKY FLIES

A fly is a little bug.
Flies fly!

There are many kinds of flies.

When we talk about flies, we mostly mean houseflies.

A black fly is very little.

A mosquito is a fly.

A fly has two wings.

A fly has six legs.

A fly lays many eggs
at one time.

Baby flies look like
little worms.

Flies can look freaky!

Words to Know

eggs

housefly

leg

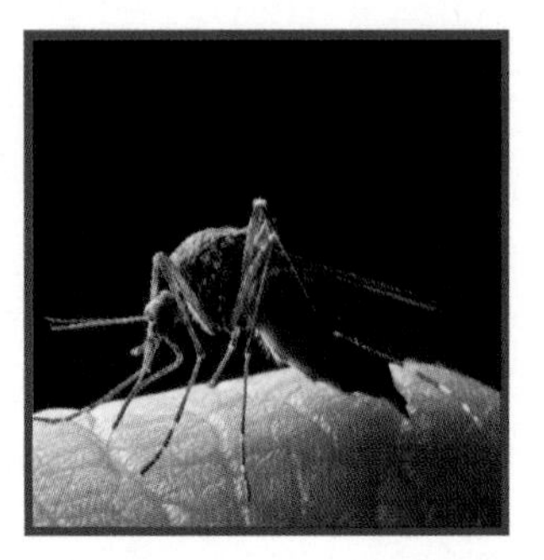

mosquito

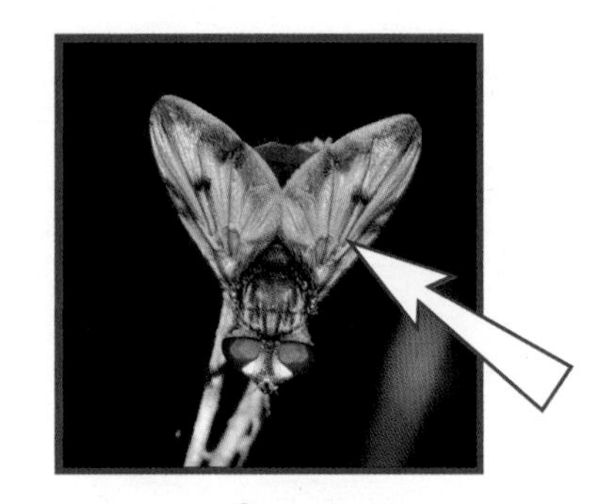

wing